Health 96

华丽的天竺葵
Gorgeous Geraniums

Gunter Pauli

冈特·鲍利 著
凯瑟琳娜·巴赫 绘
何家振 译

学林出版社
www.xuelinpress.com

丛书编委会

主　任：贾　峰

副主任：何家振　郑立明

委　员：牛玲娟　李原原　李曙东　吴建民　彭　勇
　　　　冯　缨　靳增江

丛书出版委员会

主　任：段学俭

副主任：匡志强　张　蓉

成　员：叶　刚　李晓梅　魏　来　徐雅清　田振军
　　　　蔡雾奇　程　洋

特别感谢以下热心人士对译稿润色工作的支持：

姜竹青　韩　笑　贾　芳　刘　晓　张黎立　刘之杰
高　青　周依奇　彭　江　于函玉　于　哲　单　威
姚爱静　刘　洋　高　艳　孙笑非　郑莉霞　周　蕊

目录

华丽的天竺葵	4
你知道吗？	22
想一想	26
自己动手！	27
学科知识	28
情感智慧	29
艺术	29
思维拓展	30
动手能力	30
故事灵感来自	31

Contents

Gorgeous Geraniums	4
Did you know?	22
Think about it	26
Do it yourself!	27
Academic Knowledge	28
Emotional Intelligence	29
The Arts	29
Systems: Making the Connections	30
Capacity to Implement	30
This fable is inspired by	31

在南非桌山脚下，两株天竺葵正在享受着阳光。

"你知道，三百多年前我们的祖先第一次被带到欧洲。"白色天竺葵说道。

粉色天竺葵回答："是的，从那以后我们广受人们喜爱。一些天竺葵秋天枯萎了，但春天又开始发芽。人们就是喜欢我们复生的能力！"

Two geranium plants are enjoying the sun at the foot of Table Mountain in South Africa.

"You know, our ancestors were sent to Europe for the first time more than three hundred years ago," says the white geranium.

"Yes, and we've become very popular since," answers the pink geranium. "Some of us die back in autumn and then reappear in spring. People just love us for our ability to spring back to life!"

两株天竺葵正在享受着南非的阳光

Two geranium plants in sunny South Africa

桉树吸干了我们的水

Eucalyptus trees suck up all our water

"说得对,我们中的一些人确实很怕冷。我们需要充足的阳光,不能有太多树荫。待在这些高山硬叶灌木林中就挺好,这里几乎没有树木,恰好适合我们生存。"

"几乎没有树木?你没看见附近种了那么多桉树吗?它们把我们的水全吸干了!"

"You're right, some of us do suffer from the cold weather. We need a lot of sun and not too much shade, so we do well living in the fynbos, where there are hardly any trees. That suits us just fine."

"Hardly any trees? Haven't you seen how many eucalyptus trees have been planted around here? They suck up all our water?"

"嗯，是的，我看见了。人类是多么没远见啊！荷兰人把我们带到荷兰，然后英国人又把澳大利亚的桉树带到这里！"

"荷兰人把我们带到欧洲是因为我们的花很美，而且种类丰富。"

"对，不过我们还得有其他让人喜欢的理由！"

"Oh yes, I have. How short-sighted of people to do that! The Dutch took us to Holland, but then the Brits brought Australian trees here!"

"The Dutch took us to Europe because we have such a wide variety of beautiful flowers."

"That's right, but surely we should be liked for other reasons as well!"

我们还得有其他让人喜欢的理由！

We should be liked for other reasons as well!

呼吸问题

Problems with their breathing

"是的！非洲人，如科伊桑人、科萨人，他们在这里居住了很久很久，久到没人记得他们是什么时候来的，他们从来不只把我们当花瓶。他们知道我们的价值。"

"开普省的很多人，特别是年轻人，都患有鼻塞或者呼吸问题，你知道吗？"

"So true! African people, like the Khoisan and the Xhosa, who have lived here for as long as anyone can remember, never look at us for our beauty only. They know that we are good for them."

"Did you know that many people, especially young people, here in the Cape have blocked noses or problems with their breathing?"

"鼻塞？他们不知道我们南非天竺葵能够治疗感冒吗？我们还能帮助治疗支气管炎甚至扁桃体炎呢。"

"什么是扁桃体炎？"

"嗯，扁桃体是人喉咙后边的两个淋巴腺。它们是人们抵御呼吸系统疾病的第一道防线，需要保持完好才能保护人类的健康。如果受到感染，人就会得扁桃体炎。"

"Blocked noses? Don't they know that us Southern African geraniums can help cure colds? We are also good for bronchitis too and can even cure tonsillitis."

"What is tonsillitis?"

"Well, tonsils are two lymph glands people have at the back of their throats. As they are the first line of defence against disease, they need to be in great shape to protect a person's health. It is when these become infected, that a person gets tonsillitis."

能帮助治疗感冒和扁桃体炎

Can help cure colds and tonsillitis

用我们的叶子泡茶喝，能够减轻症状

Make tea from our leaves, it will offer relief

"这是不是说扁桃体在防止人生病的时候，它们自己也生病了？"

"恐怕是这样。如果孩子患了扁桃体炎，在急着送医院之前，家长可以用我们天竺葵的叶子泡茶给孩子喝，能减缓症状。这是最好的方法，因为我们的叶子是天然的药物，而且还免费呢！"

"Does that mean that when they prevent a person from getting sick, these tonsils get sick themselves?"

"I am afraid so. But there is something parents can do before rushing a child with tonsillitis to the doctor. If they make tea from our leaves, it will offer relief. It's the best, as it is natural and free!"

"我听说我们细小的油腺一年能产两到三次油,这种油能杀死使人类生病的细菌。"

"是的,还不止这些呢。你知道吗?我们的几滴油就能使人的头发变得美丽、闪亮,而且我们的油对干性皮肤也很好。"

"真神奇!我听说,人类还把我们用在食物中。我们能让食物变得更美味可口。"

"I've heard that we produce oil from our tiny oil glands two or three times a year and that this oil can destroy the microbes that make people sick."

"Yes, and that's not all. Did you know that a few drops of our oil give people beautiful, shiny hair? And that it also helps for dry skin?"

"Incredible! And they can use us in food too, I hear. We provide fragrance and flavour."

还把我们用在食物中

Can use us for food too

独特的非洲风味

Unique African taste

"就这么简单。只需要在把我们的叶子放进罐子里,加上橄榄油,放到阳光下晒两星期,然后加到沙拉里。我们提供了独特的非洲风味,可以与最好的地中海美食媲美。"

"完全正确。所以,如果只看重我们美丽的外表,岂不是很可悲吗?"

"And it is so easy. You simply put our leaves in a jar with olive oil, leave it in the sun for two weeks, and then add it to a salad. We provide a unique African taste that can compete with the best Mediterranean cuisine."

"Exactly. So, isn't it a pity that we are usually judged only on our looks?"

"就是！我们被某些人称为老鹳草，而另一些人称我们天竺葵，太可悲了，不同的叫法引起了很多混乱。更糟糕的是，在大洋的彼岸，我们被称为鹤嘴！"

"对我来说怎么称呼并不重要，只要我能以最好的方式为大家服务就好。我得为世界做点什么。"

……这仅仅是开始！……

"Absolutely. And it's also a pity that we are called geraniums by some and pelargoniums by others, which causes some confusion. And to top it all, on the other side of the ocean, we are called crane's-bills!"

"It doesn't matter to me what I'm called, as long as I can be used in the best way for all I have to offer the world."

… AND IT HAS ONLY JUST BEGUN!…

……这仅仅是开始！……

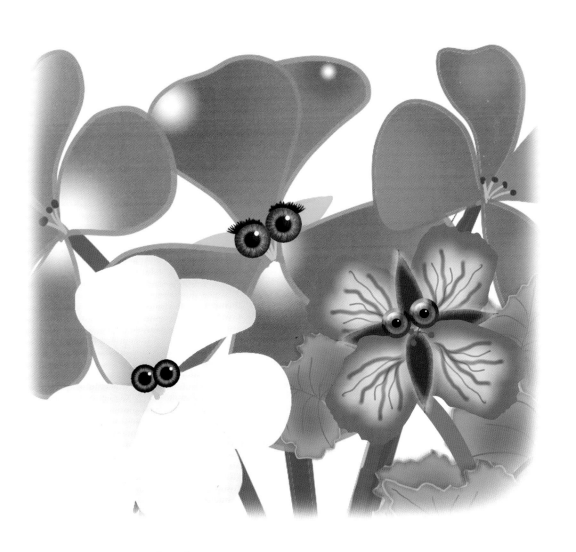

...AND IT HAS ONLY JUST BEGUN!...

Did You Know?
你知道吗？

Pelargonium plants are used to treat bronchitis (as it has an antibiotic effect) and as an immune system stimulant. It is a natural remedy used by Zulu, Xhosa, and Khoisan people.

天竺葵被用来治疗支气管炎（因为它有抗菌作用），还可以被用作免疫系统的增强剂。它是祖鲁人、科萨人、科伊桑人使用的一种天然药物。

Pelargonium root extract has anti-bacterial properties (preventing and curing bacterial infections) and anti-viral properties (preventing and curing viral infections), and is an expectorant (aiding in clearing mucus from the airways).

天竺葵根的提取物具有抗菌属性（预防和治疗细菌感染）和抗病毒属性（预防和治疗病毒感染），而且是一种化痰剂（帮助清除呼吸系统的粘液）。

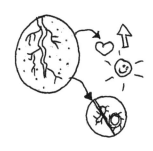

天竺葵这个名字起源于古希腊语单词"pelargós",意思是"鹳",因为其果实看起来像鹳的嘴。天竺葵的原产地是非洲南部和西南部,在其280个亚种中,90%是在这里发现的。剩下的10%是在澳大利亚发现的。

The name Pelargonium is derived from the Greek word pelargós, meaning 'stork', as its seedpod looks like a stork's bill. The Pelargonium is native to southern Africa and south-western Africa where 90% of the 280 varieties are found. The other 10% are found in Australia.

天竺葵因为其美丽和芳香而被种植。天竺葵的叶子和花可以食用,可被用于制作蛋糕、果冻和茶,其花瓣还可以用作天然颜料。

Pelargonium species are grown for their beauty and fragrance. The leaves and flowers are edible and are used in cakes, jellies, and teas. The petals can also be used as a natural pigment.

老鹳草通过弹射传播种子，弹射距离可达 10 米。老鹳草的这种能力来源于其果实的形状，它的果实形状像鹤的长嘴。另外一些种类的天竺葵羽状的荚果能在微风中漂浮 100 米。

Geraniums disperse their seeds by catapulting them up to 10 m away. This is made possible by the shape of the seedpod, which resembles a crane's long bill. Other Pelargonium species have feathered seedpods that allow seeds to float on a breeze for up to 100 m.

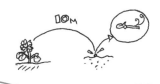

喀拉哈里沙漠、纳马夸兰、卡鲁的原始居民是以采集为生的桑族人和以放牧为生的科伊人。他们也居住在南开普省和西开普省的海岸地区。他们利用天竺葵的营养价值和医疗作用为自己服务。

The original inhabitants of the Kalahari Desert, Namaqualand, and the Karoo were the foraging San people and the pastoral Khoi people. They also inhabited the southern and western Cape coasts. They made use of the Pelargonium for its nutritional value and health benefits.

在班图语中，有非常精确的词汇为不同种类的天竺葵属植物命名。

In the Xhosa language there are very precise words such as *umtetebu*, *umuncwane*, *ibhosis*, and *iyeza lesikhalih* for naming different pelargonium species.

天竺葵属植物被顶端长有油腺的细绒毛覆盖，这些油腺发出很强的气味。天竺葵的气味有很多种，如橙子、玫瑰、杏仁、苹果、椰子、柠檬、肉豆蔻、薄荷、草莓，甚至还有一种被叫做"火味"。

Pelargonium plants are covered in fine hairs that have oil glands at the end, producing strong scents. Scented geraniums range from orange to rose, almond, apple, coconut, lemon, nutmeg, peppermint, and strawberry scents, and there is even one called "fire".

Think About It
想一想

If you get sick, would you prefer to use antibiotics from a pharmaceutical company or would you first try natural products that have been used successfully for millennia?

如果你生病了，你愿意用药厂制造的抗生素，还是先尝试一下千百年来已被证明有效的天然药物？

你认为为什么人们会给同一种植物起不同的名字，如老鹳草、天竺葵和鹤嘴？

Why do you think people give the same plants different names, as has happened with the geranium, Pelargonium, and crane's-bill?

Do you like people based on their beauty, personality, friendship, or workmanship?

你喜欢一个人，是因为他（她）外表好看，还是因为他（她）的品格、手艺或者你们之间的友谊？

你愿意别人仅以外表评判你吗？

Would you like to be judged by your looks only?

Do It Yourself!
自己动手！

Pelargonium, geranium, and crane's-bill flowers come in a variety of different colours. Nearly all these plants grow in one small region of the planet, a unique floral kingdom of the Western Cape Province in South Africa called fynbos. See how many different geranium colours you can identify by doing an Internet search for images of these plants.

天竺葵属植物的花有各种各样的颜色。几乎所有天竺葵都生长在地球上一个很小的区域，这是南非西开普省内一个独特的植物王国，叫作南非高山硬叶灌木林。上网搜索一下天竺葵的图片，看看天竺葵的花有多少种颜色。

TEACHER AND PARENT GUIDE

学科知识
Academic Knowledge

生物学	天竺葵属植物的果实是有五瓣双悬果的分裂果；天竺葵属植物被有油腺的细绒毛覆盖，这些绒毛主要用于防卫，因为香味可以阻止食草动物；桉树需要大量的水。
化 学	天竺葵属植物的花瓣含有使君子氨酸——一种能够通过干扰其神经递质使甲壳虫麻痹的化学物质；花葵苷（天竺葵宁）是一种取自猩红色天竺葵属植物的颜料。
物 理	带"羽毛"的天竺葵属植物的种子成熟时会从荚中喷射出来，在太阳照射下变干；荚的各部分的连结组织以不同的速度收缩，导致荚如同受到弹簧作用一样突然崩裂，将种子弹射出去。
工程学	为了达到最远的水平喷射距离，应该以45度角度发射。
经济学	天竺葵属植物的经济价值远远超过仅仅作为花卉的观赏价值；天竺葵生长在干燥的环境里，只需要很少的水维护。
伦理学	喜欢某人是因为他的长相或者他是谁以及他能为我们的幸福作什么贡献；世界各地的天竺葵属植物一般都是从南非引进的，但是南非人从天竺葵身上得到的经济利益远低于其他地区的人；人们对外来物种对当地生物多样性的破坏作用缺乏了解，而且很多人在知道其不利影响后，也不会做任何事消除不良后果。
历 史	天竺葵属植物在16世纪90年代被引进荷兰。
地 理	纳马夸兰高山硬叶灌木林是南非西开普地区一个独特的生物圈。
数 学	伽利略算出当一个抛射物体垂直运动时，重力拉动物体以每秒9.8米的加速度落向地面，而水平运动则保持不变，从而建立了精确的数学曲线，称为抛物线。
生活方式	欧洲城市在举行最美花卉比赛时，使用最多的就是天竺葵；空气污染主要是由柴油发动机产生的烟气和颗粒物造成的。
社会学	在远古时期，人们用几种不同颜色的天竺葵来测试色盲；科伊桑语变得越来越濒危，目前只有25万人说这种语言。
心理学	天竺葵属植物整个夏季都开花，不需要园丁们付出太大的努力，就能为人们提供美景和快乐；天竺葵的美丽对人们的情绪健康有积极效果，并且有利于增进亲密关系；天竺葵的花香能促进社交互动。
系统论	植物不仅是我们共享的生态系统的一部分，也是我们物质、精神和情绪健康的重要组成部分。

教师与家长指南

情感智慧
Emotional Intelligence

天竺葵

天竺葵表现了对自身历史的了解。他们知道自己倍受欢迎的原因是不需要太多照料，耐干旱，还能提供美丽的花朵。他们也知道自身的不足：怕冷，依赖阳光。天竺葵承认荷兰人在将他们从南非运往欧洲并进行交易的过程中所起的作用。他们也承认英国人引进外来树种（如桉树）对生物多样性具有消极影响。天竺葵质疑他们的价值只限于"颜值"，提醒大家，南非地区传统的居民非常珍视天竺葵的多重价值。现在，天竺葵已经准备好让人类重新发现他们独特的优势。他们认为名称混乱并不要紧，真正重要的是因为他们为世界提供了美好，并且受到了赞赏。

艺术
The Arts

天竺葵的花瓣不仅可以吃，而且缤纷多彩。我们烘焙一个蛋糕吧，用天竺葵的花瓣装点蛋糕！把花瓣放在蛋糕酥皮上，做成有趣的图案。比如，你可以从做一个笑脸开始。然后试着做更复杂的图案，或许可以做一朵五彩的花。使用天然的材料来赞美自然，然后，你就可以享用一个漂亮、味美的蛋糕了。

TEACHER AND PARENT GUIDE

思维拓展
Systems: Making the Connections

南非西开普地区具有独特的生物多样性。纳马夸兰高山硬叶灌木林是一个绝无仅有的植物王国和独特的生态系统，以拥有种类繁多的开花植物为傲，如今却因为外来生物（特别是桉树）的引进受到压力。按照重商主义的逻辑，桉树生长快，能为建筑和燃料提供充足的木材，因而被引进到这里。但是，桉树的引进给非洲南端脆弱的生物圈带来了严重破坏。严峻的现实是这种树的耗水量超过了生态系统的供给能力，剥夺了本地物种生存所必需的少量水分。桉树的种植在给当地环境带来压力的同时，也扼杀了与自然的生物多样性密切相关的真正的经济潜力。首先，这个地区独特的植物生物多样性只剩下了一种功能，即树木和花卉的出口。这种状况已经持续了三个世纪。其次，土著人的文化传统，包括科伊桑语和班图语，提供了数千年来积累下来的关于纳马夸兰高山硬叶灌木林中动植物知识的宝库。植物为人类提供了食物和药材，但是人类将植物的贡献降低到仅仅是好看的花朵。这使植物不能充分发挥潜力。南非西开普省失业率很高，虽然有如何利用植物的知识，但是迄今为止这些知识还没被转化成任何经济利益。

动手能力
Capacity to Implement

我们需要了解植物对人类作出的特殊的贡献。查阅百科全书，列出天竺葵属植物的用途清单，通过网络搜索进行验证。这份清单将会很长。在获得总体印象后，识别出所有经过现代科学检验的医疗应用。"经过科学验证"是指，这些应用已经完成安慰剂对照研究。在这种研究中，研究者给一组被试用一种中性物质（安慰剂），把要测试的疗法实施在另一组被试身上。对比两组的测试结果，看看被测试的疗法是否比安慰剂更有效。在你的清单中有多少种天竺葵具有所声称的效果呢？检查一下那些已经经过科学验证的天竺葵属植物，是否能在当地商店里买到它们的提取物。或许，这里面有你意想不到的机遇！

教师与家长指南

故事灵感来自

特瑞达·普雷克尔
Truida Prekel

特瑞达·普雷克尔生于南非比勒陀利亚。她在南非大学获得商务领导力硕士学位，并从事妇女管理潜能领域的开创性研究。她在鼓励妇女在职业领域承担领导责任，并促进管理的创新和改革。她的志愿工作涵盖了广泛的领域，从为社会变革而进行的音乐教育到社区安全和扫盲教育。特瑞达在勒娜特·库切的经典著作《自然的盛宴》的出版工作中起了关键作用，这本书歌颂了早期人类的饮食文化传统以及曾经与自然环境和谐相处的科伊桑人。它是同类书中第一本，而且是用石头纸印刷的。

更多资讯

https://www.teachengineering.org/view_lesson.php?url=collection/cub_/lessons/cub_catapult/cub_catapult_lesson01.xml

http://www.entente-florale.eu/results-2015/

http://www.livescience.com/14635-impression-smell-thoughts-behavior-flowers.html

图书在版编目（CIP）数据

华丽的天竺葵：汉英对照 ／（比）冈特·鲍利著；
（哥伦）凯瑟琳娜·巴赫绘；何家振译. -- 上海：学林
出版社，2016.6
（冈特生态童书. 第三辑）
ISBN 978-7-5486-1061-8

Ⅰ. ①华… Ⅱ. ①冈… ②凯… ③何… Ⅲ. ①生态环
境－环境保护－儿童读物－汉、英 Ⅳ. ① X171.1-49

中国版本图书馆 CIP 数据核字（2016）第 125789 号

————————————————————————

© 2015 Gunter Pauli
著作权合同登记号 图字 09-2016-309 号

冈特生态童书
华丽的天竺葵

作　　者——	冈特·鲍利
译　　者——	何家振
策　　划——	匡志强
责任编辑——	匡志强　蔡雪奇
装帧设计——	魏　来
出　　版——	上海世纪出版股份有限公司 学林出版社
	地　址：上海钦州南路 81 号　　电话／传真：021-64515005
	网　址：www.xuelinpress.com
发　　行——	上海世纪出版股份有限公司发行中心
	（上海福建中路 193 号　网址：www.ewen.co）
印　　刷——	上海丽佳制版印刷有限公司
开　　本——	710×1020　1/16
印　　张——	2
字　　数——	5 万
版　　次——	2016 年 6 月第 1 版
	2016 年 6 月第 1 次印刷
书　　号——	ISBN 978-7-5486-1061-8/G·396
定　　价——	10.00 元

（如发生印刷、装订质量问题，读者可向工厂调换）